# Molding & Candle Making

*by*

**Richard Taylor**

A guide to the refinement and use of beeswax in candle making, completely illustrated.

# BEESWAX

Molding & Candle Making



ISBN 978-1-908904-10-2

Published by Northern Bee Books, 2012
Scout Bottom Farm
Mytholmroyd
Hebden Bridge
HX7 5JS (UK)

Design and artwork
D&P Design and Print
Worcestershire

Printed by Lightning Source, UK

# PREFACE

This booklet deals exclusively with beeswax-how to melt and refine it, its special properties as a wax, how to mold it, and the various ways it can be used for candle making. I have not discussed mineral or other waxes and their uses, as there already exists a considerable amount of literature on these, nor have I discussed candle making generally.

All of the ideas suggested here have grown from my own experience as a beekeeper and candle maker over the past several decades. Some of them appeared in my "Bee Talks" in *Gleanings in Bee Culture*, published by A. I. Root Company, Medina, Ohio, in 1972 and 1973. This booklet greatly expands upon the methods described there, however, and I have also been able to incorporate here many more pictures. I express my thanks to the editors of *Gleanings* for permission to reproduce some of that material here.

R.T.

Richard Taylor, 1919-2003, really had two parts to his interesting life. He received his PhD in Philosophy and became a professor at three prestigious universities in the United States and a visiting lecturer at others. In that field he was a noted author of several books and numerous papers. The other part of his life was devoted to beekeeping. He was well known for producing prize-winning comb honey and was much in demand to give presentations on both beekeeping in general and his method of producing comb honey. At beekeeper meetings he was always glad to share his knowledge of honey bees and also how he produced his famous comb honey. Richard began writing articles for *Gleanings in Bee Culture*, today known as *Bee Culture*, in 1966. His regular monthly articles began in 1970 and continued until 2000. Two of his beekeeping books, *The How-to-Do-It Book of Beekeeping* and *The New Comb Honey Book*, are out-of-print, but can be found in the used book market. However, his other two books, *Joys of Beekeeping* and *Beeswax Molding and Candle Making*, are available. Over the years some people became beekeepers after reading *Joys*; all who read this book find a new enjoyment in keeping bees.

Ann Harman 2012

# BEESWAX

Pure beeswax, properly salvaged from honeycombs and honeycomb cappings and uncontaminated with resins, propolis or other impurities, is one of the most interesting and beautiful products of nature. Though it physically resembles other waxes in conspicuous ways, it is in fact entirely different in its chemistry and its molecular characteristics. Its melting point is high, between 143 and 1450 F. Mr. H. H. Root, in the leading encyclopedia of apiculture, * notes that beeswax, for its ductility, has the highest melting point of any wax known. This partly accounts for its great value as a candle wax. Beeswax candles burn very slowly, and hence are probably worth their extra cost in terms of this alone. Beeswax is extremely stable, subject to deterioration by very few solvents. It has even been brought up from shipwrecks after decades under the sea, still in good condition. The ancients thought that beeswax was gathered by the bees from plants, and Aristotle recorded this as fact. He doubtless mistook pollen for beeswax, or perhaps he confused it with propolis, a resinous pitch which the bees do in fact gather from plants. In fact, of course, beeswax is derived from the same source as honey, namely, from the nectar of flowers, and is essentially a further elaboration of honey itself. The comb building bees, having engorged themselves with honey, hang quietly in festoons within the hive, and after several hours they begin to secrete tiny white flakes of: wax from special glands beneath their abdomens. These flakes, which resemble snowflakes, are then molded and sculpted by the same bees into the beautiful and complex honeycomb that forms the basic physical structure of their city.

Beeswax is of course very valuable, usually worth far more than the honey which is the main objective of beekeeping, but produced in far lesser amounts. A beekeeper normally produces several tons of honey in the course of producing, as a by-product, a hundred pounds of wax. Systematic beekeepers try to save bits and pieces of wax while working around their hives, and by the end of the season these increase to an impressive and valuable accumulation. Candle companies are always ready to pay high prices for crude or unrefined beeswax, the demand being sustained, in part, by traditions which require them in ecclesiastical services. Persons who are not beekeepers can usually purchase wax at great savings from beekeepers, since it is an inevitable by-product of their craft. It will usually need clarifying, but this is easy enough to do and the saving makes it worth the small effort. The best wax comes from the cappings the beekeeper scrapes from the combs

*THE A.B.C. AND X.Y.Z. OF BEE CULTURE, pub. by the A. I. Root Co., Medina Ohio.

before spinning the honey out of them, and is accordingly called cappings wax. This is the major source of all domestically produced beeswax. Cappings wax is not only prettier, but also harder than the wax from combs and scraps, since it is less admixed with resins. Ideally beeswax should be a bright, lemon yellow, though it seldom is. The major pollutant of beeswax is propolis. This substance, which is ubiquitously present in bee hives, is not a wax, and beekeepers should make every effort to keep it out of beeswax, especially cappings. The practice of so many beekeepers of throwing all scrapings of wax and propolis into the same hopper with cappings is a serious mistake, causing a debasement of an otherwise precious wax. Once beeswax has been melted there is no way of removing propolis from it.

The other most common cause of the discoloration of beeswax is melting it in containers of iron, aluminum, copper and certain other metals. The only common metals which do not react adversely with beeswax to darken and discolor it are stainless steel and properly tinned cans. Since the latter are very common and cheap, however, there is no reason, apart from ignorance, why wax should be debased from this cause.

Beeswax, of whatever quality, always develops, fairly quickly, a dull, powdery surface, called the "bloom." I have never understood the reason for this, or why it should be so called, but it is not a sign of deterioration. It is in fact quite attractive. It can in any case be removed easily with a few strokes of a cloth, restoring the wax to a brilliant shine.

There are few beekeepers who are not fascinated by beeswax, and thankful that their craft has this aspect. The wax is useful, interesting to work with, and above all, singularly beautiful. Blocks of bright, clean, nicely molded wax are a pleasure to look at and quite easy to produce. There are dozens of uses for them, in various crafts, but the one that is of widest interest is of course candle making. Beeswax candles are not merely novel, they are also beautiful, practical and fragrant. Candlemakers have to date not been able to produce any very good imitation of them using cheaper mineral waxes.

## BEESWAX MOLDING

I shall describe two simple methods of molding beeswax for general purposes, whether by beekeepers or others, and then, for the benefit of beekeepers, describe the refinements needed to make a perfect block of bees wax, one that will win the first premium in any honey and beeswax show. There are some tricks to this that have hitherto not been very widely known. Then I shall describe in some detail the three basic

**A primitive but useful beeswax melting outfit. Chunks of beeswax, comb, and beeswax refuse are thrown into a tub of boiling water, and the melted wax is then ladled into paper milk cartons with an old saucepan.**

ways of making beeswax candles- by dipping, molding and rolling. Each method has its advantages and disadvantages, but they have in common that the product of each is, in its own way, lovely to behold.

The basic principle underlying beeswax molding is that beeswax is light and floats easily on water. Virtually all the impurities normally found in beeswax, on the other hand, and particularly honey, are heavy. Therefore when the wax is melted in hot water it rises to the surface while impurities settle out below.

The simplest method of molding beeswax is therefore to throw it into a tub of hot water, along with the dirt, dead bees, and other impurities that are admixed with it, and when it has melted, ladle it out into paper milk cartons with an old sauce pan. The wax blocks thus obtained, with minimal effort, still contain impurities, especially on the bottom surface, but they are free from honey and the color is usually quite good. These blocks will command the top price at any candle factory or bee supply outlet. They can also be given a further refinement, by the method about to be described, yielding blocks that are absolutely pure and suitable for the most demanding and fastidious purposes.

Here must be entered a word of caution. Beeswax is quite flammable. One must accordingly be very careful, when melting it in water over a burner, not to let the water rise to boiling temperature. If steam forms underneath the layer of molten wax it can cause the wax to rise and boil over quite suddenly, and if this hot wax flows into the burner

**Blocks of crude beeswax, suitable for dipping candles but not for molding them**

**A simple home-made beeswax melter that refines the wax to the utmost purity. Elevated container is a hot water reservoir with adjustable spigot. Lower container is melter proper with spout, from which clean molten wax flows into a mold, here shown as a small plastic bucket.**

below there is danger of violent fire. There is really no great danger here if one keeps an eye on things, so as to reduce the heat or turn it off entirely as soon as the wax begins to show signs of turbulence. The heat needed to melt the wax is considerably below the danger point anyway.

To get wax blocks of almost perfect purity one needs only to modify the primitive method just described, constructing a simple wax melter. The wax melter is made from an ordinary five-gallon can, of the kind beekeepers refer to as a "sixty", because it holds sixty pounds of honey and is in common use by commercial beekeepers. Remove the top of this can with a can opener or chisel. Next solder a spout in the middle of one side about an inch or two from the top. The best and simplest spout consists of nothing more than a tiny frozen juice can with top and bottom removed. Having soldered this on, poke a hole through the side of the sixty, through the spout, in case this was not done before the spout was soldered on.

There you have your melter; that is all there is to it. To use it, proceed as follows. Set this melter, with a few inches of water in it, over a gas burner, and get it boiling. It is best to use soft water, or rain water, but this is not essential. The water should also be acidulous, so if you think of it, add a few spoonfuls of vinegar; but this too is not essential. When the water is boiling well begin throwing in wax scraps, or partially refined blocks, such as those obtained by the crude method described above. As the wax melts down add more, stirring from time to time, and trying to keep the water just below boiling, so as to get the wax melted as quickly as possible but without risking danger of its going over the side. Eventually the melter will be filled, nearly to the spout, with a deep layer of molten wax. Beneath this layer will be a resinous mass of residue which beekeepers refer to graphically as "slumgum," consisting of impurities that have settled out of the wax, and beneath this, of course, the original bottom layer of hot water. Now, when everything is melted, the burner is turned off, For awhile there will be a few globs of slum gum floating about on the surface of the hot wax, as well as a few dead bees that have floated up. Most of this material will gradually settle out and go to the bottom as the wax begins to cool a bit, and as one gives it an occasional stir with a wire. When everything has settled, but before the wax begins to congeal around the edges of the melter, start pouring very hot water into the top of the melter. As this water goes straight to the bottom, the wax rises and flows out the spout and into molds. A bit of flyscreen fixed over the edge of the melter, on the inside, will hold back any dead bees or other impurities still floating on the wax.

The most natural way to introduce hot water into this melter is of course with a tea kettle or large pan, kept filled and hot over a nearby burner. This has the disadvantage that the hot water is introduced irregularly, as the container is moved back and forth between burner and melter, so that the wax flows from the spout irregularly too. Moreover, one hand of the operator is occupied in pouring water when he really needs both hands to replace wax molds as they become filled. A simple and worthwhile improvement is therefore to set up a hot water reservoir equipped with spigot so that one can regulate an even flow of water into the melter. Thus, solder a spigot or small valve to the bottom of another topless five-gallon can and set this, filled with hot water, over a burner that is raised above the level of the melter. The spigot is adjusted to produce a steady and even flow of hot water, producing exactly the same even flow of wax from the spout, and enabling the operator to change molds in a steady and routine fashion as they successively fill with molten wax.

By this system, using the simplest home made equipment, one can

**Small beeswax cakes molded in dixie cups.**

in a single operation turn out hundreds of identical little wax cakes, using dixie cups as molds, or just as easily mold a single large block, or several. If one is making small wax cakes in large numbers, using dixie cups or something similar as molds, he might as well pop a little candle wick into each as it begins to congeal, thereby producing, at almost no extra effort, little candles of pure beeswax having a great appeal to the eye. The candle wick should first be impregnated with wax, by dunking a few feet of it into the wax melter, and then chopped up into short pieces. Thus the wicks will remain stiff and erect when poked through the partially molten wax cakes.

Paper milk cartons make perfect wax molds and cost nothing. A quart carton makes a block weighing just two pounds. When the carton is ripped away from the hardened wax, usually the next day, there will often be found impurities embedded on the bottom of the wax, particularly if the melter was used to mold wax that was very full of dirt and refuse. These can be totally eliminated by simply repeating the process just described, that is, by running the wax through the melter once again. Even without repeating the process, however, all but the last couple of molded blocks will usually be quite clean.

When the process is completed the melter will be full of hot water, with perhaps an inch of extremely dirty beeswax on top. This layer of wax is removed as soon as it begins to congeal by running a knife around the edge. Scrape the refuse from the bottom of it and save it, for it is quite valuable, notwithstanding its apear.ance. The melter itself requires no cleaning, other than a casual rinsing.

There are of course other ways of melting wax, but I know of nothing better or simpler than this if one is trying to mold clean blocks to use in candle making or for other purposes. This system eliminates every trace of honey from the wax, for example, as many other methods do not, for the residual honey is dissolved in the water below, and stays there. The system requires no straining, other than the very rough straining of large particles or dead bees by means of flyscreen, as described; and straining is, in fact, a very ineffective way of clarifying beeswax.

## MOLDING WAX FOR SHOW

The method just described yields beeswax that is perfectly adequate for most purposes, including candle dipping, and wax that is thus run through such a melter a second time is likely to be perfectly clean, with no residual impurities whatsoever. Even wax that is discolored and contains considerable sediment is suitable for candle dipping, however, provided one also has on hand some wax of the highest quality and purity to use for the final two or three dips. Molded candles must of course be made entirely from first quality wax, and wax molding for more fastidious purposes, such as competition for premiums in fairs and honey shows, is an art in itself.

The challenge to a true beekeeping enthusiast is to produce a block of wax that will take a first premium in a highly competitive honey and wax show, of the sort sponsored each summer by the Eastern Apicultural Society (and also The National Honey Show in London). Such a block of wax has no special commercial value, that is to say, no significant value over and above what is paid for unrefined wax of good quality, but a successful response to the challenge to produce such a specimen is a reward in itself.

The first requirement here is to avoid discoloration by impurities in the wax or, what is more common, by reaction with metals, as already described. One is likely to suppose that pure beeswax should be snow white, but in fact it is not, and wax that has been bleached will actually lose points in most honey shows. Even the pure white foundation used in comb honey production turns out to be yellow if melted down. The true color of beeswax is, as noted, lemon yellow, and any departure from this is a fault.

As beeswax cools and solidifies it contracts, and this is what produces the greatest problems. Hence, if the cooling is not slow and uniform, the block cracks. Small cakes have less tendency to crack than large ones, and very small ones will almost never crack if made in paper molds. In the case of a very large cake of wax, however, the center will still be liquid for some time after the sides have solidified, and as the center then cools, congeals and contracts the surface becomes very vulnerable to cracking. A block of cracked wax is automatically eliminated from consideration in a competitive show, though of course it is perfectly satisfactory for all other purposes.

The way to prevent cracking is to let the wax cool slowly in a warm room. If paper molds with flexible sides, such as milk cartons, are not used, then the sides must be greased or coated with very soapy water so that the wax will not cling to the mold and inevitably crack. In case one is molding a very large block of wax, weighing five pounds or more, then even further precautions are needed. The simplest is to place the mold filled with molten wax in a corrugated carton, cover this with another such carton, for insulation, then cover the whole works with blankets, and give it a couple of days to cooL I have found this system unfailing even with the largest molds, such as a plastic heart-shaped waste basket used to mold the beeswax heart illustrated.

Even though cracking is thus avoided, however, the wax shrinks in volume as it congeals, and there is no way under the sun to prevent this. In case one uses a round mold this does not particularly matter, for the resulting block of wax, though somewhat shrunken in size, is still round, and thus of uniform shape. It therefore becomes a challenge to produce a

**A large prize winning block of pure beeswax molded in a heart-shaped plastic waste basket.**

fair sized brick or cube of wax, uniform in shape, and weighing at least two pounds, preferably five pounds or more. This is so difficult, to get a truly uniform block with straight edges and flat surfaces, that most exhibitors settle for round blocks molded in round pans. These present no great difficulty and are acceptable in any show, but they do not really stand out as anything exceptional. This brings us to the real secret of wax molding.

Let us suppose you are using a square mold, such as a milk carton. A full gallon size is excellent, as this can be used to make an almost perfect cube of an impressive size. The difficulty is, however, that if the wax hardens without cracking, then the resulting block has sides that are concave and edges that are not straight, and this is aesthetically displeasing. It is as though the sides had been sucked in, which is in effect just what happens. The larger the volume of wax the more pronounced will be this effect. It is a real problem, but it has a simple solution, as follows.

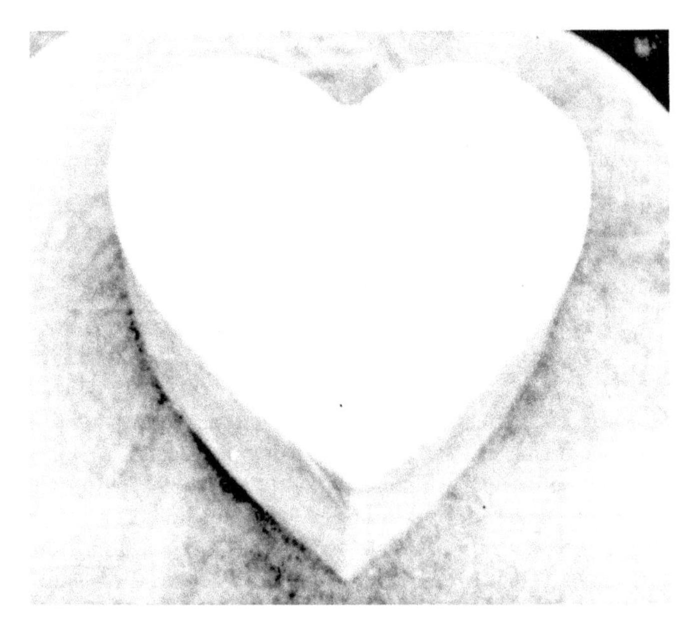

**The same wax heart fresh from the mold, before the surfaces were made smooth by the method of fry pan and foil.**

Spread aluminum foil over the bottom and up the sides of an electric frying pan or, lacking this, a regular frying pan of ample size, and set the temperature to a low heat, sufficient to melt beeswax. Now lay the block of wax in the pan and slide it gently about. The edges melt down slowly and uniformly until that side is perfectly flat. If the wax tends to stick to the foil, then increase the heat a bit. Pour off the melted wax from time to time, and repeat this operation with each of the remaining five sides. The result is a perfectly uniform cube or brick of wax, achieved in this simple and direct way.

**A large prize winning block of pure beeswax molded in a heart-shaped plastic waste basket.**

## BEESWAX CANDLE DIPPING: EQUIPMENT.

The object of the candle dipping system I shall now describe is to get a large number-- a hundred or more-- identical candles by a simple and routine operation and without waste of wax. The secret of it is a set of simple home made dipping frames, one frame for each ten candles desired. And the general principle is that wicking is strung on these frames, which are then dipped into the wax in succession until a fair thickness of wax has built up on the wicks. This ensures perfectly straight and uniform candles. The frames are then dismantled and the dipping continued, five candles at a time, until the candles are finished.

**Large and small dipping cans with ash can which is used as water bath.**

I shall first describe the equipment needed, with precise directions for making up the special frames, and then describe the procedure. Meanwhile, however, what are the advantages of dipping? Why hand dip candles when they can be made in simple molds? There are several advantages. For one thing, the typical mold makes only six candles at a time. With the dipping system, on the other hand, one can make however many he pleases and has simple frames for. The frames cost almost nothing, while molds are expensive. The wax used for dipped candles need not be the best. It can be dark and discolored and it will not matter, so long as the last two or three finishing dips are made in wax of the very best quality. Moreover, hand dipped candles are something special, and are generally so regarded, especially - if made of beeswax. They are thus quite rightly considered the product of a special craftsmanship. And of course the dipping is fun. To go round and round, frame after frame, and see a hundred identical candles take shape, is a rather unique satisfaction. No wax is wasted, for what is left at the end of the operation is all set to be used next time,

Now for the equipment needed, all of which is -extremely simple,

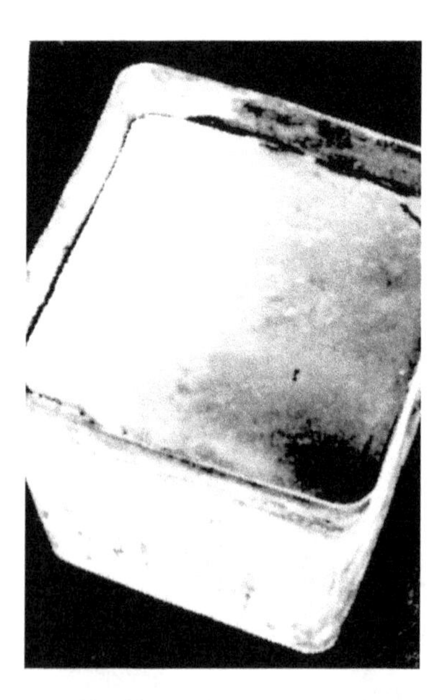

**The dipping cans seen from above, each containing congealed beeswax from the previous dipping operation. Metal shoulders on the small dipping can are without significance.**

1. Two dipping cans. One of these is a five-gallon can with the top cut off. The other is the same depth and length, but not as wide. The large one holds melted wax of whatever quality one happens to have and is used for all but the last two or three dips. The smaller one holds melted wax of the very finest quality and is used for the final finishing dips. It should be 14 x 9 1/2 (like the larger one) but only three inches wide.

2. A water bath can. This is nothing but a galvanized can or tub, about the same height as the five-gallon dipping can. The dipping cans, one at a time, are set in this and surrounded by hot water, and the whole business set over a low burner, to keep the wax melted. Since beeswax is lighter than water, the dipping can must be held down with wires at the edges. Pieces of a wire coat hanger, the ends bent to make hooks, work fine. Put a few little slats of wood under the dipping can, so water can circulate underneath.

3. Candle wicking. This is not just wrapping cord, which will not work, but woven wicking made especially for candles. It can be purchased at hobby shops, candle factories, or from the A. 1. Root Company in Medina, Ohio, or from Betterbee, Inc., RR #4, Box 4070, Greenwich, NY 12834. Specify the diameter (7/8 “) and kind (beeswax) of candle the wicking is for. That is considered a standard diameter, but I find that a 3/4” diameter at the base is more pleasing to my eye.

**Receptacles for melting wax to replenish dipping cans, bundle of wicking and snippers.**

4. Various receptacles and pitchers for melting wax and replenishing the dipping can as one- proceeds. I use a couple of topless maple syrup cans and a stainless steel pitcher I found someplace. Do not use copper, iron or aluminum. These receptacles are submerged in a pan of hot water over a low burner to melt the wax as one goes along.

5. A cool room. A cool basement is perfect if one has a stove with at least three burners down there. The room should also have beams overhead with twenty or more cup hooks screwed about a foot apart. Lacking these, one can string a few lengths of knotted rope overhead. This is where the dipping frames hang between dips.

6. A few odds and ends such as pocketknife, scissors and heavy rubber bands.

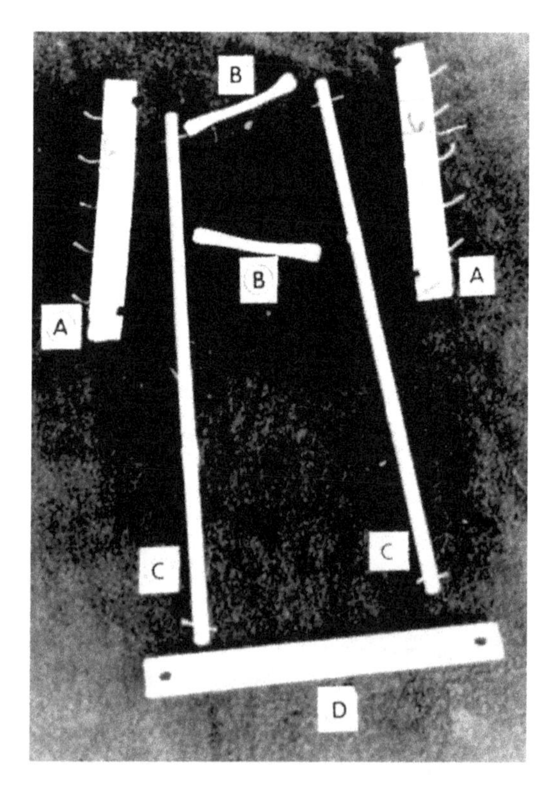


**Unassembled frame, showing (a) two halves of numbered top bar with five headless nails along each side and cup hook handle in center of each, (b) two heavy rubber bands, (c) side bars consisting of two dowels with tiny brad at each end, and (d) bottom bar with hole at each end.**

7. And finally, the dipping frames. I use ten clearly numbered frames to make one hundred candles at each operation.

A dipping frame consists of a top bar, bottom bar, and a dowel rod for each side. It is made up as follows.

The top bar is made from a piece of standard (3/4") thickness scrap wood, 8" long and about 2" wide. Bore a 1/4" hole in the center near each end, about 3/8" in from the end. Then saw it in two, lengthwise, right through the centers of the two holes. Drive five nails along each side, evenly spaced, cut their heads off, and bend them slightly upwards. The wicks will tie onto these. Then finally, screw a cup hook into the center top of each half.

The bottom bar is a strip of wood, such as might be cut from a yardstick, same length as the top but only about an eighth-inch thick and an inch wide, with a quarter-inch hole at each end. This bottom bar, unlike the top, is not ripped in two.

Each side consists of a length of quarter-inch dowel 15½" long, which can be picked up at any lumber supply. Bore a tiny hole at each end, 3/4" from one end (the top) and 1/8" from the other end (the bottom). A small brad is pushed into each of these, holes, so as to protrude from each side of the dowel.

Each half of each top bar should be identically numbered with a

**The assembled dipping frame, from above. Split top bar is held together at each end with heavy rubber bands. Wicking .is strung for ten candles.**

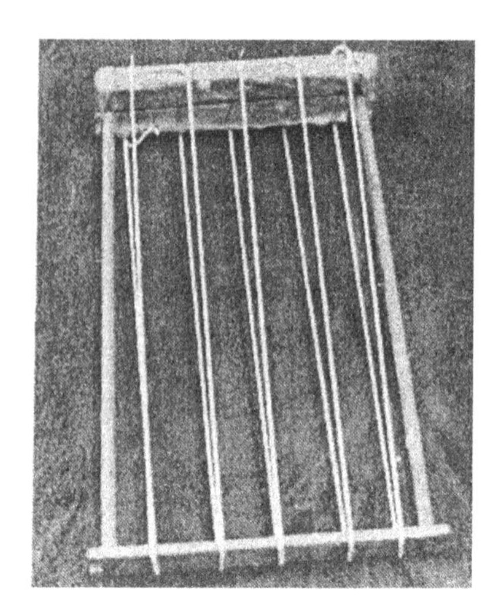

**Dipping frame from above [left] and from below [rightl. Note cup hook handles, and the manner in which tiny brads in side bars keep top and bottom bars properly separated by preventing slippage.**

crayon, so one can always tell which half goes with which. These numbers also serve to keep the several frames in proper order as the dipping proceeds, so they should be large and clear.

The frame is then assembled by pushing the dowel rods into the holes of the top and bottom bars. A heavy rubber band is wrapped tightly around each end of the top bar, to hold the two halves together.

Now string the wicking up and down, making a simple knot around each protruding nail. Pull them fairly tight, so they will be straight. The tiny brads in the dowel ends prevent the top and bottom bars from sliding along the dowels when the wicking is pulled tight. There will, of course, be ten wicks on each frame when the stringing is done.

Now, assuming you have a full five gallons of beeswax melted in the dipping can, you are ready to begin. As noted, this wax need not be of good quality, nor does it even need to be entirely free of sediment. The details of the dipping procedure I shall describe shortly, but the general idea is probably clear already. The frames are plunged into the molten wax, one after another, and wax builds up on the wicks, which are held perfectly straight by the frame. After a few dips, when the candles have grown to about the size of a pencil, the frames are dismantled by removing the rubber bands from the top bars and snipping the wicks at the bottom bars. Five partially completed candles now hang free from each half top bar, and the dipping continues as before, five candles at each dip.

## THE DIPPING PROCEDURE

The details of the procedure just sketched are as follows.

Fill the larger of the two dipping cans, the five-gallon one, with chunks of beeswax. Set it in the water bath over a low burner, adding more wax as the melting progresses. When the dipping can is filled almost to the top with melted wax, the dipping is ready to begin. Melting all this wax takes time, of course, but you will meanwhile have been stringing the wicking on the frames. You should also start melting one or two of your auxiliary cans or pitchers of wax for replenishing the dipping can as the wax gets used, up.

Now begin dipping. Pop three or four of the frames into the dipping can all the way down, and leave them there for a few minutes, *for the first dipping only*, so the hot wax will soak into the wicks. Repeat this with each frame until all the wicks are impregnated with wax. Then, each wick having become thoroughly waxed, proceed in a smooth and deliberate way to dip each frame, one after another, into the wax, all the way to the top, and remove at once. Do not (after the first dip) leave the frame in the hot wax, for this would only cause the wax to melt off the candles instead of being added to them, And do not yank the frames out too fast, or the candles will not shape up properly. The idea is simply a regular, smooth and deliberate motion. plunging each frame in all the way and then removing it at once, Hang each frame from a beam overhead after each dip and repeat with the next frame. Keep things in order, dipping frame number one. then number two, then three, and so on. Repeat the dippings in the same order after each complete round. Each dip should be clear up to the top bar. Do not make any half dips in order to get the tapered candle shape. This shape will result automatically and uniformly from full dips.

After the sixth or seventh dip the wax will have built up to about the size of a pencil. The frame itself will also be layered with wax, of course, but that is all right, Since the frames have served their purpose of keeping the wicks straight they can now be dismantled, as follows. Cut each sick loose at the bottom, with strong shears or a wire cutter. Remove the heavy rubber bands from the ends of the top bar. The bottom bar and side rods of the frame, covered with wax, now drop off in one piece and the top bar divides into two halves. From each half there now dangle five unfinished candles, about the thickness of a pencil. So what you now have are twenty holders (each of the ten original top bars divided in two), each with five half - finished candles. These are dipped in succession, as before, and hung from their cup hook handles as the dipping proceeds.

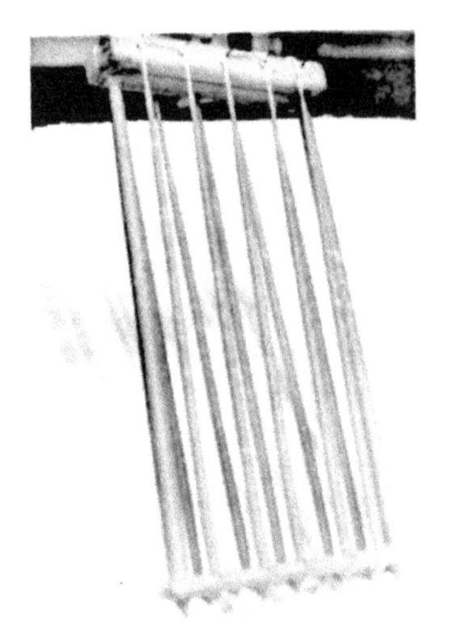

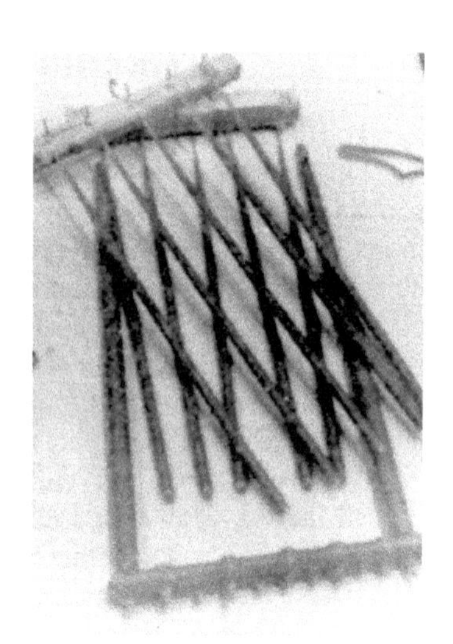

Dipping frame after seven dips. Partially finished candles are perfectly straight and about pencil size. The frame has become coated with wax from the top bar down.

After about the seventh dip the candles are cut loose at their bases and the rubber bands are removed from the top bar, allowing the top bar to split into two and the rest of the frame to fall away.

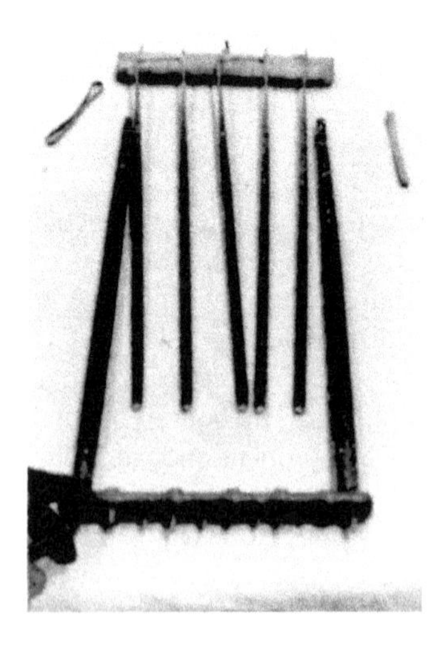

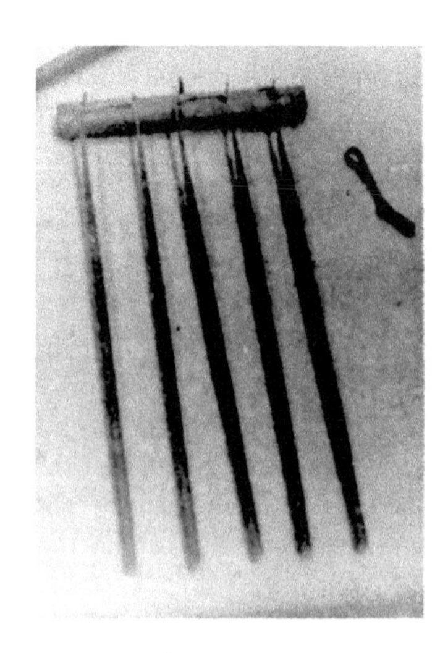

The frame, being no longer needed, is set aside, and dipping continues five candles at a time.

**Five candles needing only one more dip. Note that icicles have been trimmed from the bottoms of the three on the left but not yet from the remaining two.**

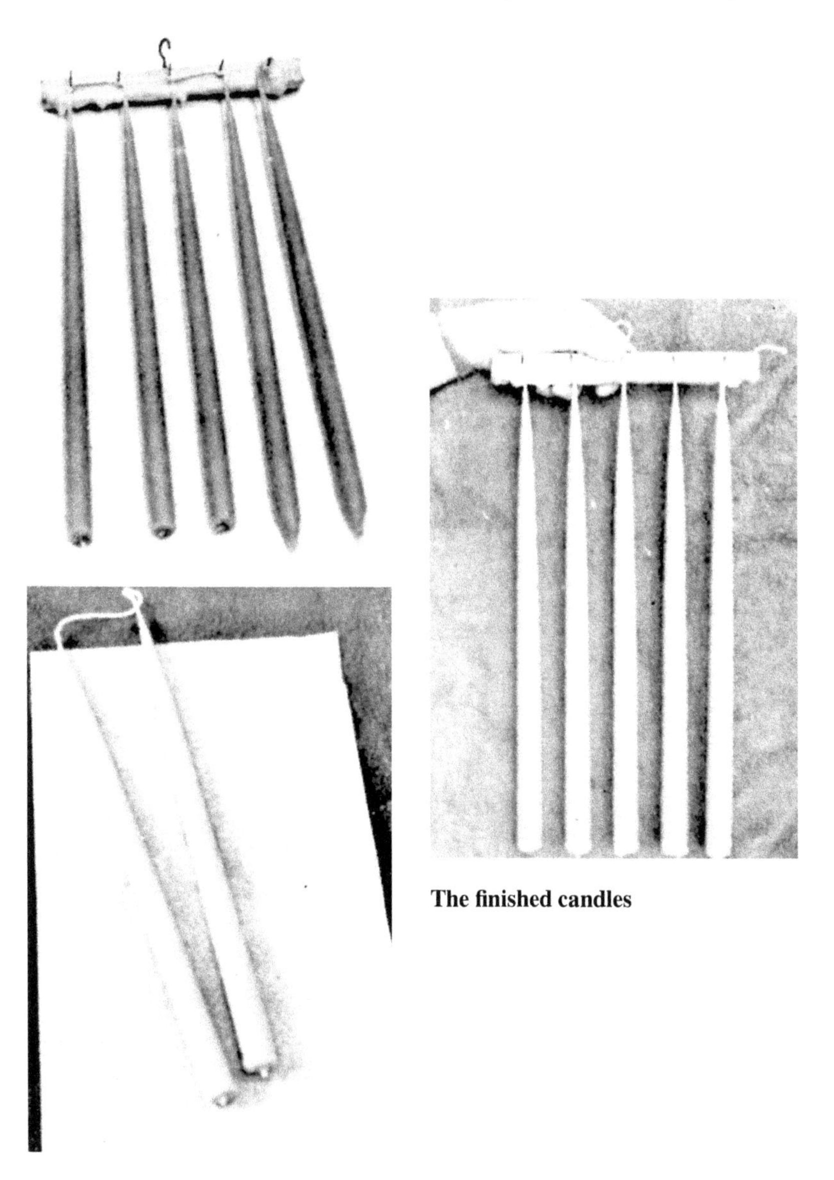

**The finished candles**

**A pair of finished candles. Note that the wick is continuous between them, and that a single drop of wax remains on the bottom of each.**

An icicle of wax develops at the bottom of each candle, and this must be trimmed off from time to time with a knife so that the candles can be plunged clear to the bottom of the dipping can. When only a few dips remain to be done, switch to the smaller dipping can which contains only the finest wax. This can, as noted, is much smaller than the original five-gallon dipping can, thus requiring less wax to fill it, although it is just as deep. When only one dip remains, which will be after about the fourteenth or fifteenth dip from when you began, trim the icicles for the last time from the bottoms of the candles. The finished candle, 3/4" to 7 /8" in diameter at the base, will thus have a drop of wax at the bottom resulting from the final dip. This is appealing and unmistakably marks it as hand - dipped. Just before the final dip one may also, if you wishe, stick a tiny label near the bottom of the candle, which will show through the last layer of wax. My own little label, put on every other candle, reads simply "100% Pure Bees wax." Now all that remains is to slip the candles from the headless nails and wrap each pair in a sheet of saran wrap or thin paper.

Both dipping cans can be left to cool with the wax in them, and they will be all set to use again next time, without the waste of a single drop. The frames are of course wax coated, but they are easily cleaned by dipping them in hot water.

The candles, if made according to the foregoing directions, are all nearly identical, weigh about five ounces per pair and measure about fourteen inches. They burn much longer than ordinary candles and are therefore probably worth their extra cost, simply in terms of utility alone. They are readily wholesaled through gift shops, or sold directly for about a dollar and a half or two dollars per pair.

**One hundred finished candles hanging from the beams where they were hung after each dipping.**

## MOLDING BEESWAX CANDLES

The basic procedure for candle molding is so obvious that extensive instructions are unnecessary.

Candles can be molded in all sorts of interesting shapes, and hobby shops offer a great variety of molds for this purpose. It must be remembered, however, that beeswax contracts considerably in solidifying, and the finished candle will therefore not correspond exactly with the mold. If the mold has square edges then the finished candle must be given final shape by the method described above for molding beeswax blocks. Shrinkage and distortion increase with the size of the mold.

Since the uniqueness and value of beeswax candles lies in their substance (beeswax), rather than in their form, there is really little point in seeking novel shapes. One should instead make attractive candles of conventional tapered shape, having in mind that it would be difficult to improve upon this shape, and impossible to improve upon beeswax as its substance.

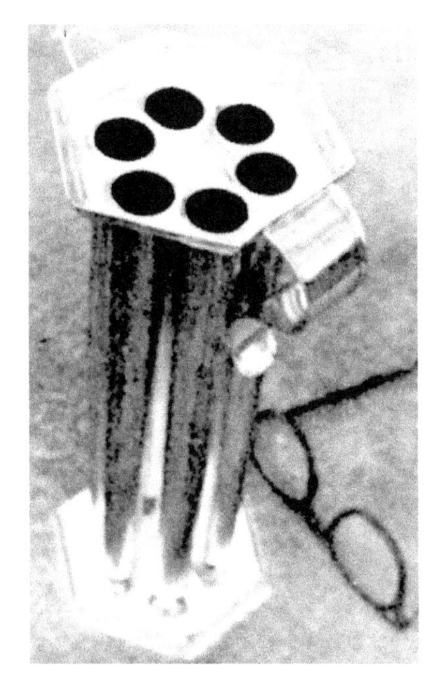

**Candle molds. The mold on, the left has the hexagonal shape associated with honeycomb and was designed to appeal to beekeepers.**

Candle molds can be purchased from hobby shops. Antique molds are also all right, provided they are not dented. Practical molds can also be purchased from the A. I. Root Company, Medina, Ohio. And E.H. Thorne, Wragby

The basic procedure for molding candles is as follows.

Melt two or three pounds of the finest beeswax in a double boiler. Never heat it directly over a burner of any kind. While the wax is melting, cut lengths of wicking such that each length is about four or five inches longer than double the height of your candle mold. Then string each such length down through one cylinder of the mold, through the hole in the bottom, up through the hole of an adjoining cylinder and out the end. Fasten the ends with a long pin, such that the pin goes through the wick and lies crosswise of the candle mold. If pins are not available, the wicks can be tied to small sticks that lie crosswise. The wicks should be pulled

**The mold ready for pouring. The wicks can be held straight and centered with sticks, as here, or more conveniently with long hat pins.**

fairly tight, and centered. Next press a pinch of putty over each hole in the bottom of the mold, so wax will not flow out. Alternatively, one can set the candle mold on a piece of flat sponge that has been dipped in cold water. This will cause the wax to congeal there and not leak out. Now pour the wax into each cylinder, fairly quickly so as to fill each cylinder in a single motion. As the wax begins to harden it will also contract, so more wax will need to be poured in; otherwise the bases of the candles will be deformed or missing entirely. As soon as the wax appears to have hardened, but while it is still warm, all excess wax in the top of the mold can be scraped out with a knife. This is a simple operation if done while the wax is still warm and soft. Now remove the pinches of putty, or the wet sponge, and the long pins or sticks, and snip the wicks between the

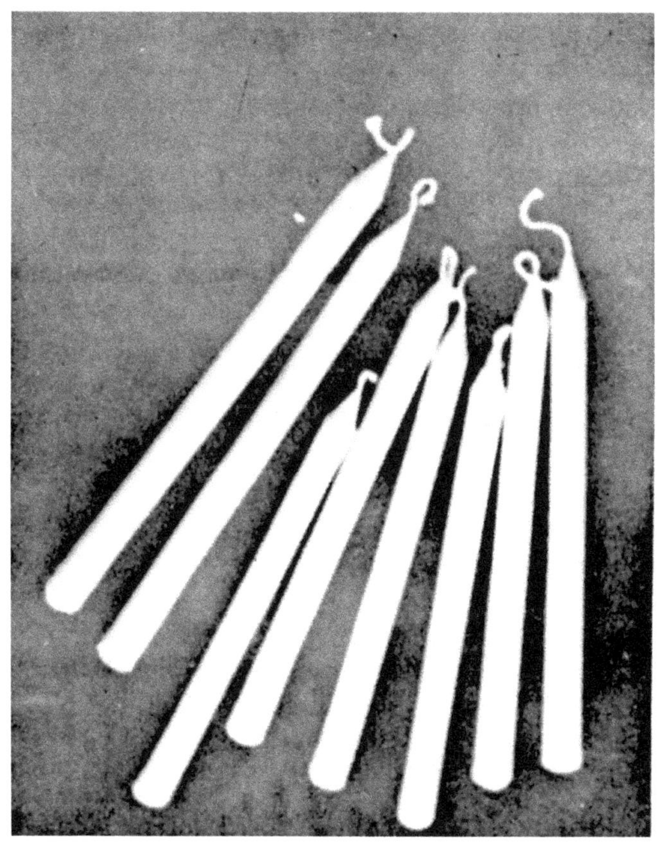

**Beautiful molded candles of pure beeswax.**

candle tips, so that the candles can be pulled from the mold. Let the wax become completely hard, then pull the candles out by grasping the wicks that protrude from the bases of the candles.

The main problem in the foregoing procedure is getting the candles out. They sometimes stick, so that no amount of strength can dislodge them. Unless, too, the wax has completely hardened, the wicks will sometimes pull out, leaving the wax in the mold. To overcome this difficulty one can put the mold in the freezer of a refrigerator before pulling out the candles. A further step in insuring that the candles will come out is to coat the inside of the mold with a very thin layer of vegetable oil just before introducing the hot wax. Vegetable oil can be purchased in spray cans, and this works perfectly, provided it is not allowed to remain in the mold very long and become gummy before the wax is poured in. Still another step in facilitating removal of the candles is to run warm (not hot) water over the cylinders before pulling out the candles. If by extraordinary bad luck one ends up with candles damaged in the attempt to extract them and beeswax still stuck in the mold, then that beeswax will have to be entirely cleaned out before the mold can be used again. This is done by submerging it in a solution of hot lye water. Lye is about the only common solvent that will loosen beeswax. Bear in mind that hot lye is extremely caustic and dangerous, particularly to the eyes. Read the directions and precautions on the can before proceeding.

Molds make beautiful beeswax candles, of strictly uniform appearance, and although only a few can be made at one time, the setup needed is a very simple one. Instead of setting aside a half a day or more for a large operation, as with candle dipping, one can on the spur of the moment make up a half dozen candles with little work.

# CANDLE ROLLING

In the past few years it was discovered that beautiful beeswax candles could be made by simply rolling regular beeswax foundation around a wick. "Foundation" is the name for the pure beeswax sheets embossed with the pattern of the honeycomb used by all beekeepers to induce the bees to build their combs in wooden frames. Candles made this way burn nicely, though not as long as dipped and molded ones, and the embossed honeycomb design not only makes them exceptionally attractive but is very appropriate to their being made of beeswax.

Beeswax foundation sheets can be obtained from any bee supply outlet, or from any beekeeper, for beekeepers always have a supply on hand for use in their apiaries. Only "full depth" sheets are of use for candle making, that is, sheets measuring 8 ½" x 16 ¾".

At first these candles were made simply by cutting a sheet of foundation diagonally and rolling each resulting triangle around a length of wicking, but it was soon discovered that much more attractive candles could be made by shortening these triangular pieces from the pointed end. This of course immediately led imaginative persons to experiment and invent all sorts of imaginative shapes, so that today these candles can be found in an unlimited variety. Special booklets are available from hobby shops and from the Dadant Bee Supply Company in Hamilton, Illinois, illustrating some of the numberless possibilities.

These candles soon became so popular among beekeepers, and their wives, that the bee supply companies began making the foundation sheets in colors. The next step was to admix generous amounts of parafin with the beeswax, both as an economy and to permit of easier and better coloration. Today one can purchase honeycomb sheets measuring 8"x 16 5/8" in an enormous assortment of colors. They should no longer be called "foundation," however, since they are composed mostly of waxes other than beeswax and cannot be used in hives as the foundation of honeycombs. Most hobby shops stock them, and they can also be purchased from the Dadant Bee Supply Company, exclusively for candle making.

The best rolled candles are made by combining two or more colors, resulting in candles that are very striking. If one wishes to make any sort of small commercial enterprise of such candles he needs a standardized procedure, such that the candles are all of the same size and shape, differing only in color.

**Pattern for cutting honeycomb sheets for rolled candles, shown from top and bottom. Note strip of wood under lower edge for positioning the pattern at edge of table.**

By using a pattern, such as the one illustrated, one can cut two full sheets at a time, one colored and the other always white, yielding pieces for four regular candles, plus small scrap pieces that can be rolled into two small candles. If the small candles are then sold for twenty cents a pair - and they sell very readily at this price - this comes fairly close to paying for the total wax used for the large candles as well, and the price received for those, usually seventy-five cents to a dollar, is almost a net profit.

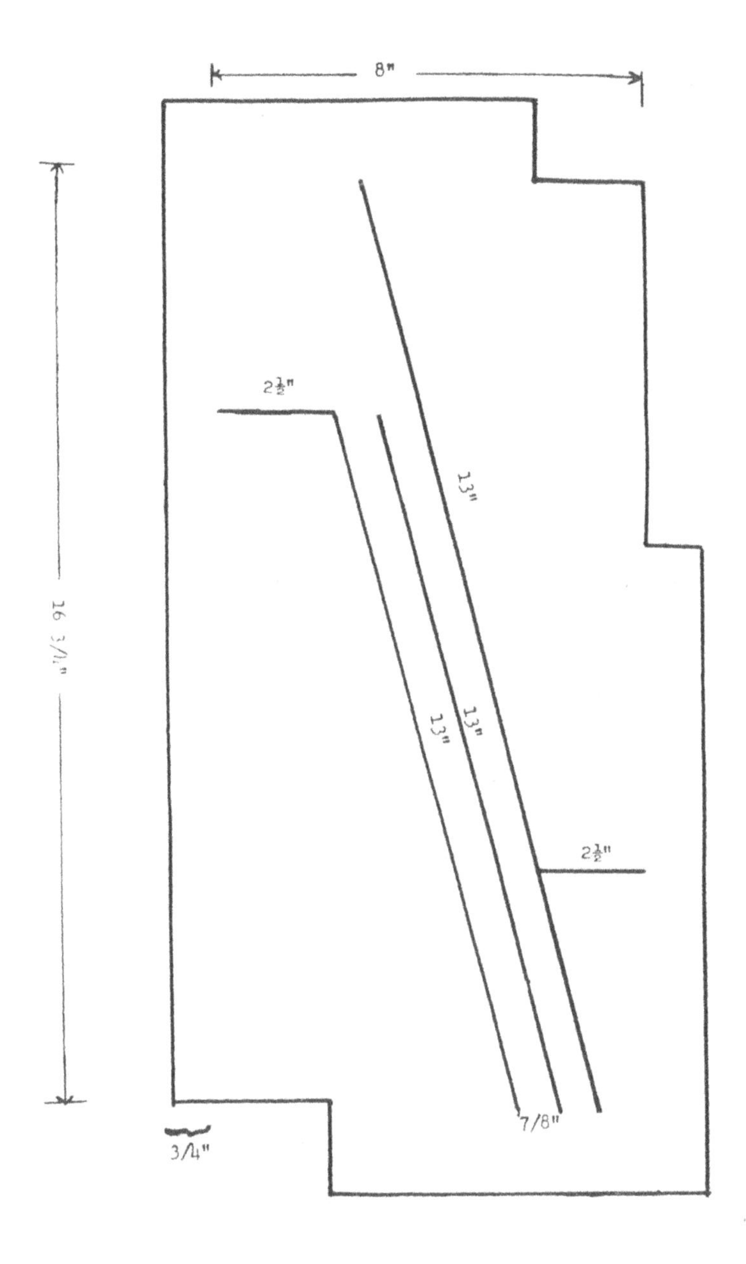


**Pattern is cut from piece measuring about 10" x 2", and a stick about ¾"wide is affixed to the bottom edge underneath, as a guide. Honeycomb sheet measuring 8" x 16 ¾" fits under pattern with no edges or corners showing.**

The pattern illustrated is the invention of Mr. Marvin Schultz, a resourceful New York State beekeeper whose wife has made a small business of candle rolling. It can be made from thin plywood, veneer wood or as mine was, from masonite. The specifications are given here, though they need not be absolutely precise. The slots are made with a saber saw.

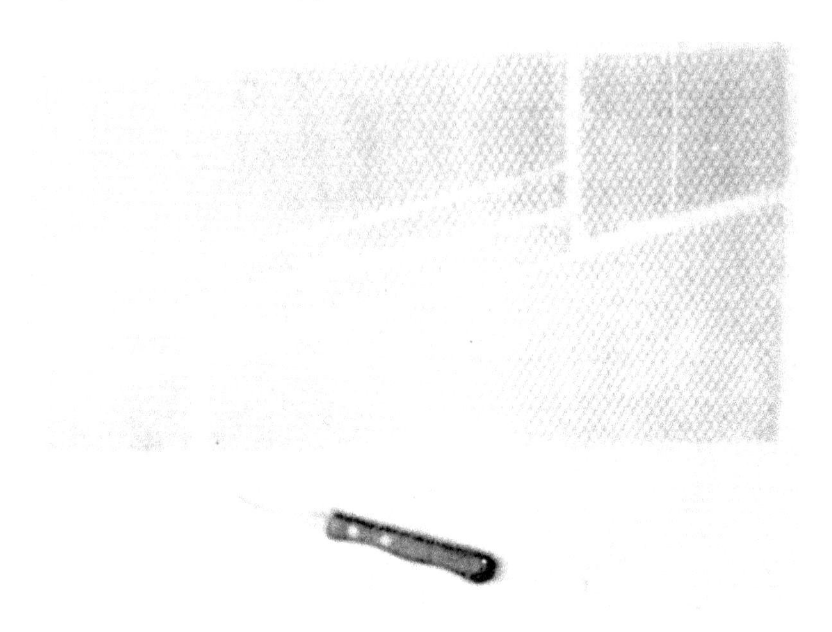

**A single honeycomb sheet cut by pattern. The piece in the lower left and that in the upper right corners are scraps, used for making small candies. Normally two sheets are cut at once, one pure white and the other colored.**

One uses this pattern as follows. Lay two full honeycomb sheets on a table or smooth board, their edges coinciding with the edge of the table or board. Lay the pattern over them, such that it just fits with no foundation protruding at any edge. A strip of wood under the bottom edge of the pattern positions it correctly on the two sheets, by fitting just over the edge of the table or board. Now simply run a knife along the slots in the pattern, and the pieces are all cut. The regular candles are made by pressing the narrow strips along the diagonal edges of the larger sheets of different color, then rolling the resulting sheet around a length of regular candle wicking. The narrow strip thus ends up as an attractive spiral on the candle. The small scraps are rolled in a similar way, to make little candles. All this must be done in a fairly warm room,

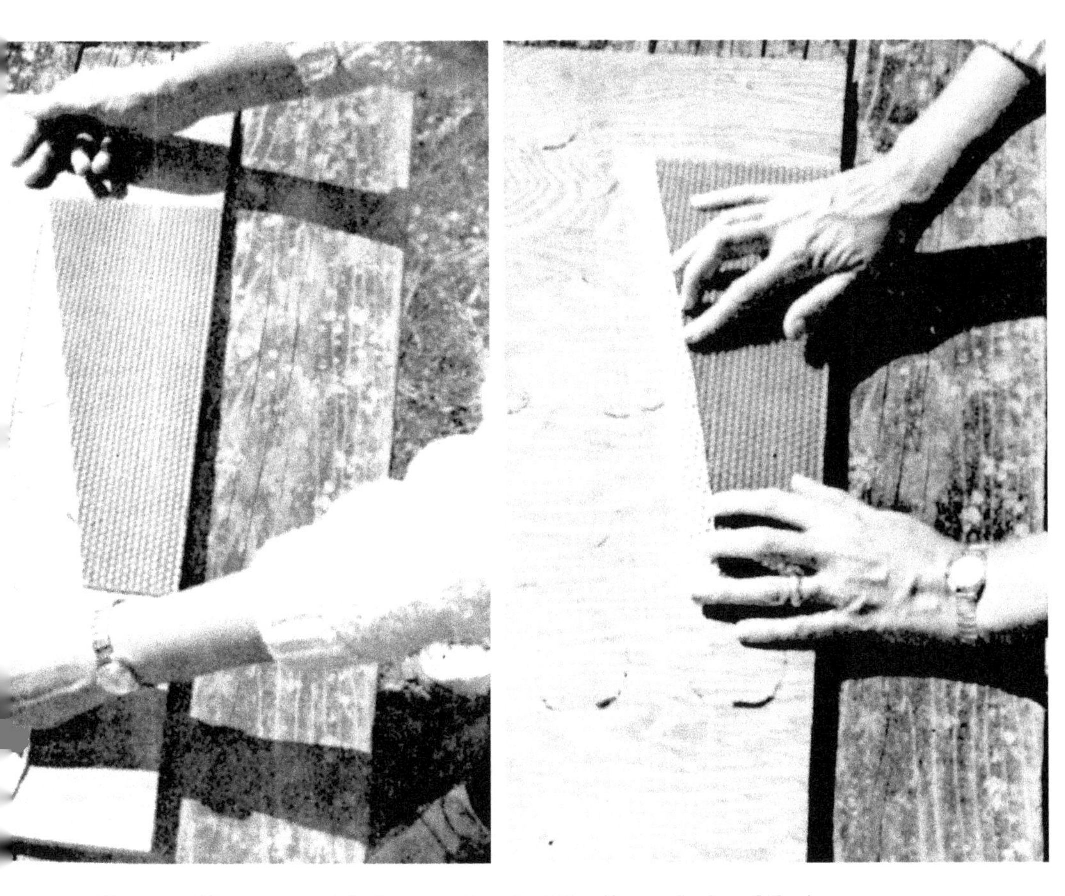

**Step one: The narrow strip is pressed against the diagonal edge of the larger piece. Nothing more is required to hold it in place.**

**Step two: The edge of the honeycomb sheet is bent over a length of wicking.**

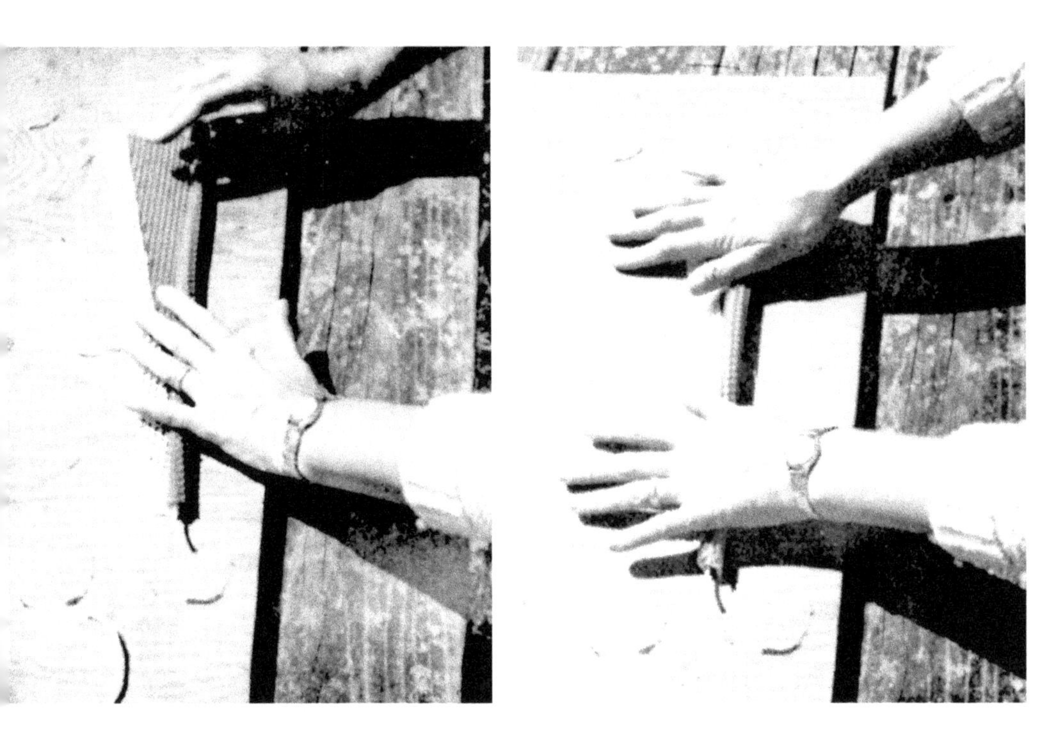

Step two: The edge of the honeycomb sheet is bent over a length of wicking.

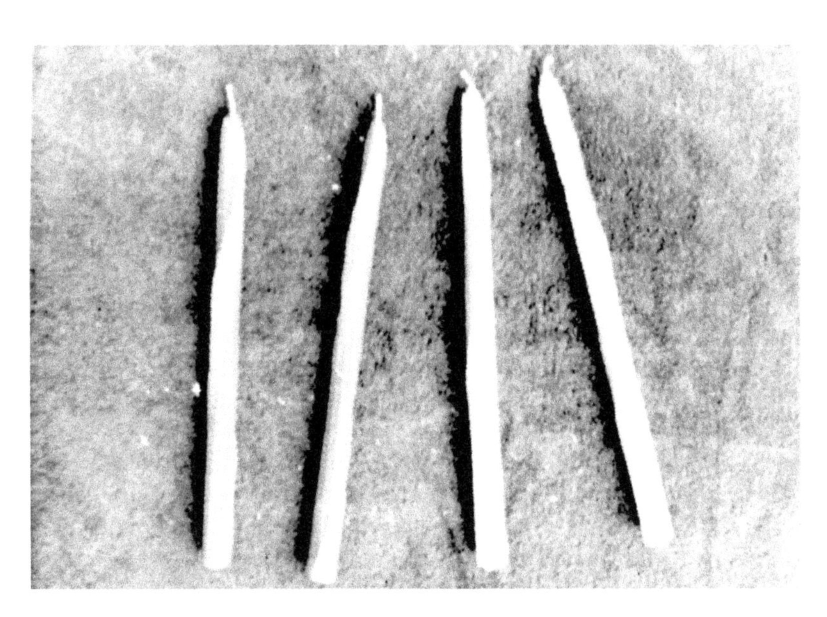

Finished candles cut from two sheets. Pair on the left is blue with white spiral trim, pair on the right is white with blue spiral trim.

**Four regular candles and four shorties, showing relative sizes.**

so that the sheets will roll easily and without stiffness. The finished candles can then be wrapped in saran wrap or plastic, with a typewritten or printed label inside. Such a label is desirable, to explain what the candles are and, particularly, that they are hand-made.

It is important that these candles be properly displayed for sale. A very effective display is to stand them upright in a large butter crock, so that the crock is quite filled with them. They thus give a pleasing impression of colorful stick candy, and have a considerable appeal. They must not, of course, be described as beeswax candles, except in the case of those that are in fact rolled from pure beeswax foundation. They are quite generally known just as "honeycomb" candles.

Lightning Source UK Ltd.
Milton Keynes UK
UKHW020650061120
372920UK00012B/532